# Principles of Earth Grounding Resistance

Created in cooperation
with Fluke Corporation

AMERICAN TECHNICAL PUBLISHERS
ORLAND PARK, ILLINOIS 60467-5756

Glen A. Mazur

American Technical Publishers, Inc., Editorial Staff

Editor in Chief:
    Jonathan F. Gosse
Vice President—Production:
    Peter A. Zurlis
Art Manager:
    Jennifer M. Hines
Digital Media Manager:
    Carl R. Hansen
Technical Editor:
    James T. Gresens

Copy Editor:
    Talia J. Lambarki
Cover Design:
    Jennifer M. Hines
Illustration/Layout:
    Thomas E. Zabinski
    Robert M. McCarthy

*The authors and publisher are grateful for the technical information and assistance provided by the following companies and organizations.*
USDA Natural Resource Conservation Services

*National Electrical Code and NEC are registered trademarks of the National Fire Protection Association, Inc.*

1 2 3 4 5 6 7 8 9 – 14 – 9 8 7 6 5 4 3 2 1

Printed in the United States of America

ISBN 978-0-8269-1436-1

 This book is printed on recycled paper.

# Contents

# Grounding Methods and Requirements

Grounding systems are installed to protect personnel, equipment, and buildings from unwanted and dangerous ground faults. Grounding categories include the grounding and bonding for personnel protection (fault protection), grounding electrode, lightning protection, and signal reference systems. Grounding systems are tested to ensure that adequate ground resistance values are present in the system. Several different methods are used to provide grounding and are based on National Electrical Code® (NEC®) requirements.

## GROUNDING

*Grounding* is a low-resistance conducting connection between electrical circuits, equipment, and the earth. Proper wiring and grounding are required in any electrical system for proper and safe equipment operation. Proper wiring requires that a system, all loads, and circuit components be properly grounded per industry standards, IEEE, and other recognized organizations, standards, guides, and recommendations. In addition to any original equipment manufacturer (OEM) requirements for proper and safe operation. Common electrical codes and standards organizations include the following:

- National Electrical Code (NEC®)
- Occupational Safety Health Administration (OSHA)
- National Fire Protection Association (NFPA)
- International Electrotechnical Commission (IEC)
- Institute of Electrical and Electronic Engineers (IEEE)

A grounding system must not only be installed correctly, but also be designed to be in service over the expected life of the electrical system and continue to properly operate even after sustaining large current faults. To ensure that a grounding system is installed properly, is in good working order, and operates for years, several tests must be performed before, during, and after the system is installed. Grounding system tests may involve taking the following voltage, current, and resistance measurements:

- system voltage
- static electricity voltage
- system current
- leakage current
- grounding system resistance
- soil resistivity
- soil pH

### TECH TIP

*Electrical systems must be connected to a grounding electrode.*

Soil resistivity measurements are taken to determine the soil resistivity, which will determine the best location for the placement of the grounding electrode, grid, or grounding system. The soil resistivity value has to be determined before a grounding system is installed. The resistance to earth of the grounding system is taken after the grounding system is installed to ensure that the grounding system does not exceed the required maximum resistance set by the NEC® and other grounding regulations and recommendations. Soil pH measurements are sometimes taken to determine which metal (copper, stainless steel, or galvanized steel) is the best material to use in a specific location.

Testing the grounding system is important during initial installation, but just as important as part of a routine preventive maintenance program. This is because a grounding system can be damaged over time due to such conditions as corrosive soil, loose electrical connections, and damaged components. Testing must be performed at regularly scheduled intervals because environmental conditions may change. For example, drying of the soil changes the moisture content and then can cause changes in the grounding system. Testing a grounding system is performed during and after system installation using earth (ground) testers and ground clamp meters. **See Figure 1-1.**

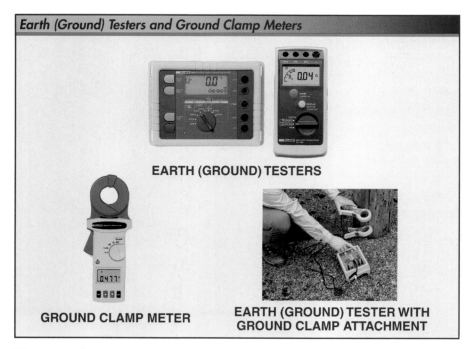

**Figure 1-1.** Ground system tests are performed during and after system installation using earth (ground) testers and ground clamp meters.

## GROUNDING SYSTEMS

The function of a grounding system, is to provide a safe path for fault current to flow. *Fault current* is any current that travels a path other than the normal operating path for which a system was designed. The proper grounding of electrical tools, machines, equipment, and distribution systems is an important factor in preventing dangerous conditions and allowing electrical and electronic equipment to operate correctly.

An overall grounding system comprises various systems or subsystems with different functions such as the grounding electrode system. The grounding electrode system provides the zero reference for the electrical system, lightning protection system, bonding system for the safety of personnel, and signal protection system. They are all interconnected with conductors and connectors to form an equipotential plane.

The conductors include the wires, connections (terminals), splices, grounding electrode (ground electrode, grid, or system), and the soil. Grounding an electrical system to earth is accomplished by connecting the grounding circuit to a metal underground electrode, the metal frame of a building, a concrete-encased electrode, a grounding ring, or other approved grounding method. **See Figure 1-2.**

### Grounding Subsystems

Each of the grounding systems enunciated above is specialized for a different purpose, and when combined, the categories provide a safe and effective grounding system for individuals and equipment. **See Figure 1-3.**

**Grounding Electronic Equipment.** Electronic equipment is grounded properly to provide a good ground for electronic systems for better communication, with less noise, with process control equipment and other systems.

A good ground to earth reduces static electrical charges, which allows signal integrity to be maintained for sensitive video, sound, data, medical, and security equipment; programmable logic controllers (PLCs); computer numerical controls (CNCs); variable-frequency drives (VFDs); and other electronic equipment. Signal reliability is difficult to maintain in electronic equipment where many signals are transmitting data at 5 V or less.

*Building ground systems can sometimes be identified by braided bare copper cable connected to copper ground rods.*

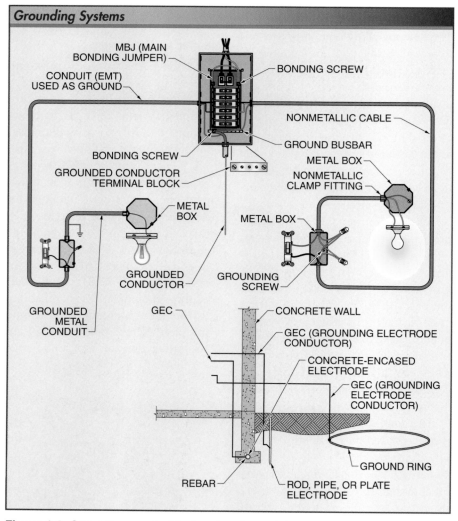

**Figure 1-2.** Grounding an electrical system to earth is accomplished by connecting the grounding circuit to a metal underground electrode, the metal frame of a building, a concrete-encased electrode, a grounding ring, or other approved grounding method.

**Bonding of Electrical Equipment.** Electrical equipment is bonded together to reduce the chance of electrical shock by grounding all exposed non-current-carrying metal. The most important reason for bonding equipment is to prevent electrical shock when a person comes in contact with electrical equipment or exposed metal.

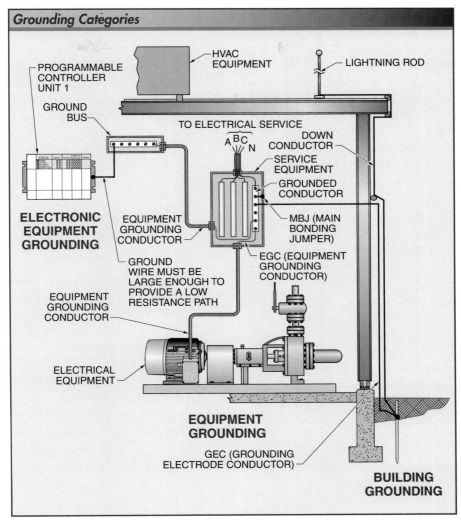

**Figure 1-3.** The three categories of grounding are electronic equipment grounding, equipment grounding, and building grounding.

Non-current-carrying metal parts that are connected to a grounding system include all metal boxes, metal raceways, enclosures, metal equipment parts, and any metal a person might make contact with that is around an electrical circuit. A fault current may exist because of insulation failure or a current-carrying wire making direct contact with a non-current-carrying metal part of a system.

In a properly grounded system, the flow of fault current must trip the overcurrent protection device (fuse or breaker). When a fuse opens or a breaker is tripped, the circuit is opened and no current flows. Equipment grounding also helps to prevent electrical shocks from static electricity and static build-up in equipment. Static electricity can also cause fires and explosions when allowed to accumulate.

**Grounding Electrode Systems.** A *grounding electrode system* is the connection of an electrical system to earth ground by using grounding electrodes, such as the metal frame of a building, concrete-encased electrodes, a ground ring, or other approved grounding method. Building grounding ensures that there is a low-impedance (low-resistance) grounding path for fault current (short to ground or lightning) to earth ground. A *low-impedance ground* is a grounding path that contains very little resistance to the flow of fault current to ground.

*Grounding electrode systems must be routinely tested with earth ground resistance meters to verify that there is a low enough resistance to provide adequate protection to equipment and personnel.*

The overall grounding system also includes a lightning protection system to protect the building, cooling tower, or outside structure from lightning strikes by providing lightning current, a path to ground. A lightning protection system must also have low-resistance because of the high current requirements created by lightning. A failure of any part of a grounding system when carrying a lightning strike significantly increases the chance of electrical flashover causing fire and building damage.

### Ground Resistance Values

There are different values given for the maximum resistance of a grounding system depending upon the reasons for grounding. For example, the electronic industry has lower resistance to ground requirements than the NEC®, for the protection of sensitive electronic equipment. The primary objective of grounding and bonding is to remove fault current as fast as possible. The NEC® states that if the grounding electrode is a single rod, pipe, or plate, it must have a resistance to ground of 25 Ω or less.

Grounding of sensitive electronic equipment is primarily designed in effect with electronic noise reduction by using a ground to eliminate noise and other unwanted induced interference, or signals. The unwanted current that is removed to ground by electronic grounding systems is typically measured in milliamperes and continues to flow as long as the electronic equipment is connected. OEMs of electronic equipment and systems typically specify grounding systems with a resistance of 5 Ω, 3 Ω, or 1 Ω or less.

A properly operating grounding system must meet the requirements and needs of the electronic, building, and electrical equipment. All grounding requirements can be met by installing a grounding system with the lowest possible resistance and by installing a system with longevity. The following are maximum resistance, or impedance, values that should be met unless specifically stated otherwise by approved agencies:

- Resistance to ground for the telecommunications and electronic industry is typically 5 Ω or less.

- Single rod, pipe, or plate grounding electrodes shall have a resistance to ground of 25 Ω or less (per NEC® requirements).

- For lightning protection grounding systems, the industry typically requires 6 Ω or less, and in areas of high incidence of lightning 1 Ω or less.

Test instruments are used to test new grounding system installations and for routine maintenance tests. Continued testing as part of a preventive maintenance program ensures that a grounding system is operating properly and safely.

## Methods of Grounding

A *grounding electrode conductor (GEC)* is a conductor that connects grounded parts of a power distribution system (equipment grounding conductors, grounded conductors, and all metal parts) to an approved grounding system. A *grounded conductor* is a conductor that has been intentionally grounded. A grounded conductor is typically the neutral conductor.

The GEC is the grounding system that provides the direct physical connection to earth. The grounding electrode is typically one or more grounded electrodes driven into the ground. The grounding electrode can also be the metal frame of a building if effectively grounded, the reinforcing bars in concrete foundations, a ground ring, a metal plate, wire mesh typically installed on rocky grades, or underground metal water pipe as long as the grounding electrode meets the low-resistance and all code requirements. **See Figure 1-4.**

---

**TECH TIP**

*Periodic maintenance should be performed on the overall grounding system to verify that the equipment grounding conductors and the grounding electrode conductor are properly sized according to the NEC®. The grounding system should be effectively grounded so that the ground fault will flow in the equipment grounding conductor to the power source and facilitate the operation of the circuit breakers.*

---

Each method of grounding has its own limits and requirements. Thus, when selecting, installing, and testing a grounding electrode, all codes and standards must be verified and implemented because not all methods are approved and work well in different locations. For example, a metal water pipe must have at least a 10′ length of pipe in direct contact with the earth and even when it does, a second separate ground rod must still be installed to meet some requirements.

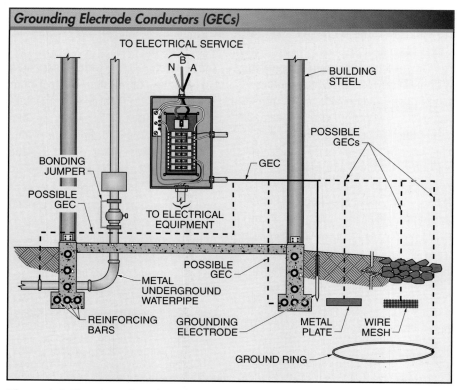

**Figure 1-4.** The grounding electrode is the grounding system that provides the direct physical connection to earth and provides the zero reference to the electrical system.

Once a method is selected and approved, the best practice is to follow the NEC® requirements, to test the system, and record all measured values during installation, during startups, and as part of a routine maintenance program. Since the NEC® does not recommend any one system but rather covers the requirements once a method is selected, it is always best to first check with the local authority having jurisdiction (AHJ) and state inspectors for requirements and practices in the given area.

# Testing Grounded Equipment and Systems

The grounding of equipment and systems protect people, equipment, and operating systems from lightning and the danger of fault current within an electrical system. Grounding is the making of an intentional connection to the ground (earth) to make the earth a conductive part of the total electrical system. To be an effective conductor of fault or lightning currents, the grounding system must have low resistance. The only method of acquiring knowledge and documentation of the grounding system and ground resistance is to use test instruments to take measurements.

## WHEN TO TEST

Ground tests must be completed during initial installation to verify that minimum resistance requirements are being met. Ground tests must be performed routinely to ensure that the system is operating properly. Older grounding systems that may have met resistance requirements at the time of installation may not meet the resistance requirements of buildings with modern, sensitive electronic equipment. Likewise, in areas of new construction, soil conditions can change due to the lowering of the water table and drying of the earth around a building's installed grounding system as more of the earth is covered with pavement and buildings. Grounding tests must be completed as follows:

• after completion of any building modifications or outside construction that may have compromised the grounding system

• once a building site is determined; a soil resistivity test must be performed

to determine the best location and type of grounding system to install

• after the grounding system is installed and before power is applied; it must be verified and documented that the grounding system meets minimum resistance requirements

• after construction is completed and the building is fully operational; it must be verified that no damage or changes have been made during construction

• once a year as part of a preventive/predictive maintenance program; the grounding system must be tested to ensure ongoing protection to personnel and equipment from electrical shock and fire

### TECH TIP

The NEC® allows a grounding electrode conductor connection to be made at accessible points at three locations: at the load end of service drop, in accessible meter enclosures, and in service disconnecting means.

## EARTH GROUND TESTS

There are various types of tests that can be performed to obtain the resistance to ground value. The most common test methods are the Wenner and the Schlumberger methods. With these methods, the ground resistance is measured, and a formula is applied to determine the resistivity value, in ohm-meters ($\Omega$m), which is performed to determine the condition of the soil in which the ground electrodes are to be installed and the best grounding method to use. This is done on new building construction sites and other structures that can carry electrical system fault currents or lightning current such as cellular telephone towers and utility substations as part of the initial design and specifications. The resistance to ground test is the actual testing of the grounding system components (rod, plate, ring, etc.) to verify and document that the selected grounding method and installation meets all minimum resistance requirements, in ohms ($\Omega$), as required by the code and industry standards. **See Figure 2-1.**

## SOIL RESISTIVITY

Determining the location, type, and size of a grounding system is usually not taken into consideration when determining the location and size of a building or structure (cellular signal towers, etc.). However, once the building or structure location, type, and size are determined, all electrical requirements are determined as part of the design.

The type and size of the earth grounding system become part of the considerations and specifications.

Soil resistivity measurements provide important information about the electrical properties (low to high resistance) of the soil at different site locations and at different depths. Since the earth grounding system is part of the total electrical system and is used to provide a safer system, properties of the earth that become current-carrying component properties must be known. However, unlike low-resistance metal conductors, soil has a much higher resistance than metal conductors and requires measurements to be taken to ensure that the lowest resistance earth grounding system is installed.

Only by taking measurements can the condition of the soil be known. Without taking measurements, one can only guess the type, size, and installation requirements of the grounding system to be installed. It is more expensive to rework an incorrectly installed system than to install it correctly the first time. Incorrect information about the grounding system can lead to the following:

- oversizing of the system in hopes that after the earth grounding system is installed, the measured value meets minimum resistance requirements (*Note:* Some oversizing is helpful, but additional oversizing only increases costs because additional materials and time are used.)
- undersizing of the system, which is expensive and time consuming because it can lead to taking shortcuts

to meet the minimum requirements that could lead to a dangerous situation at some point

- relying on false assumptions or outdated practices, such as assuming

that the resistance of the soil increases with depth and that the soil at a given site location is basically the same, so that the placement of electrodes can always be at the most convenient point

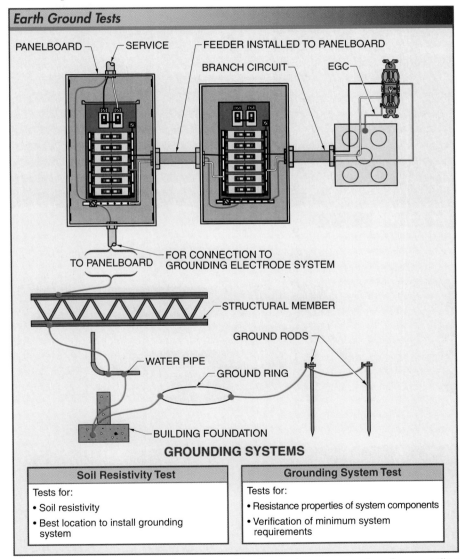

## Earth Ground Tests

**GROUNDING SYSTEMS**

| Soil Resistivity Test | Grounding System Test |
|---|---|
| Tests for: | Tests for: |
| • Soil resistivity | • Resistance properties of system components |
| • Best location to install grounding system | • Verification of minimum system requirements |

**Figure 2-1.** The resistivity test is performed for the design or expansion of a grounding system. The resistance to ground test is performed to verify compliance with code and industry standards.

The resistance of the earth (soil) always varies with soil type, moisture content, temperature, and other factors. Typically, sand and gravel conduct poorly and clay conducts much better. Although more soil moisture lowers resistance, which allows better conductivity, resistance increases when the soil freezes. **See Figure 2-2.** Understanding the type of soil and its resistance also helps in selecting the material of the grounding electrodes. In general, the lower the soil resistivity, the higher the corrosiveness of the soil. Electrodes made of stainless steel or copper (either plated or solid) are the least affected by corrosion, and galvanized electrodes are the most affected over time.

## Four-Terminal Ground Resistance Measurements

A soil resistivity test is performed to determine the best type (electrode, grid, loop, or plate) of grounding system to be used. Testing the resistivity of soil requires a test instrument such as a four-terminal ground resistance tester. The test instrument system involves four metal probes that are driven into the earth, connecting conductors long enough to connect the probes back to the meter, a tape measure, a calculator, and paper and pencils. The best type of tape measure to use is the open-reel type, which allows dirt to fall off the tape upon retracting. **See Figure 2-3.**

### Soil Resistivity

| Material | Resistivity ($\Omega$/cm) | | |
|---|---|---|---|
| | Avg | Min. | Max. |
| Fills. ashes, brine, cinders, waste, salt marsh | 2370 | 590 | 7000 |
| Clay, shale, gumbo, loam | 4060 | 340 | 16,300 |
| Soil with added sand | 15,800 | 1020 | 135,800 |
| Gravel, sand, stones with little clay or loam | 94,000 | 59,000 | 456,000 |

| Temperature* | | Resistivity* |
|---|---|---|
| °C | °F | ($\Omega$/cm) |
| 20 | 68 | 7200 |
| 10 | 50 | 9900 |
| 0 | 32 (water) | 13,800 |
| 0 | 32 (ice) | 30,000 |
| −5 | 23 | 79,000 |
| −15 | 14 | 330,000 |

* based on 15.2% moisture (sandy loam)

**Figure 2-2.** The resistivity of the soil varies with soil type, moisture content, temperature, and other factors.

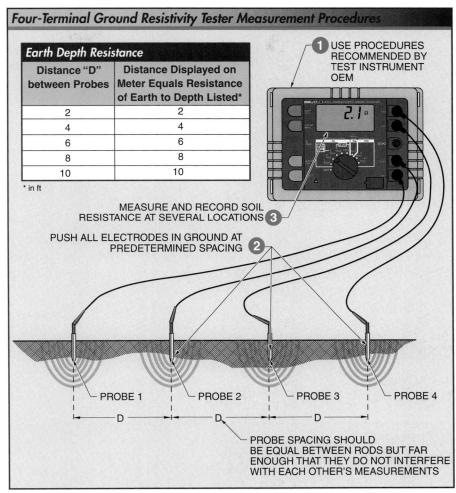

**Four-Terminal Ground Resistivity Tester Measurement Procedures**

| Earth Depth Resistance | |
|---|---|
| **Distance "D" between Probes** | **Distance Displayed on Meter Equals Resistance of Earth to Depth Listed\*** |
| 2 | 2 |
| 4 | 4 |
| 6 | 6 |
| 8 | 8 |
| 10 | 10 |

\* in ft

**①** USE PROCEDURES RECOMMENDED BY TEST INSTRUMENT OEM

MEASURE AND RECORD SOIL RESISTANCE AT SEVERAL LOCATIONS **③**

PUSH ALL ELECTRODES IN GROUND AT PREDETERMINED SPACING **②**

PROBE 1    PROBE 2    PROBE 3    PROBE 4

|← D →|← D →|← D →|

PROBE SPACING SHOULD BE EQUAL BETWEEN RODS BUT FAR ENOUGH THAT THEY DO NOT INTERFERE WITH EACH OTHER'S MEASUREMENTS

**Figure 2-3.** When measuring ground resistance, a four-terminal ground resistance tester does not use the grounding electrode as one of the probes. The tester is used to measure the resistance of soil only, not the resistance of the soil and grounding electrode.

When testing the resistivity of the soil, or earth, using a four-terminal ground resistance tester, apply the following procedure:

1. Use the recommended measurement procedure from the test instrument OEM for measuring earth resistance.

2. Push or drive all four probes (rods 1 to 4) into the ground at depth and spacing recommended by OEM. Spacing determines depth of resistance measurements. For example, when distance between probes is 6′, the resistance measurement displayed on tester is resistance of earth to depth of 6′.

3. Measure and record the resistance measurement of the earth at several different distances (4', 6', and 8' apart). Any change in resistance measurements indicates that the soil conditions are changing over the measured surface or distance.

Most ground resistance testers display readings in ohms ($\Omega$). To convert the ohm readings to ohm-meters ($\Omega$m), apply the following formula:

$$P = 2\pi \times A \times R$$

where

$P$ = soil resistivity (in $\Omega$m)

$2\pi = 2 \times 3.14 = 6.28$

$A$ = probe distance spacing (in m)

$R$ = resistance measured on meter (in $\Omega$)

*Note*: To convert feet to meters, multiply feet by 0.3048. To convert meters to feet, divide meters by 0.3048.

Test rods can be rotated 90° to get a more accurate indication of the entire soil condition. If the test results vary widely, additional measurements can be taken at 45° to give a clearer picture of the soil conditions. **See Figure 2-4.**

## GROUNDING SYSTEM TESTING METHODS

There are four methods of measuring the resistance of a grounding system. Each method has advantages and disadvantages. Understanding each measurement type allows a qualified person to select the best method for a given grounding application. The four methods of measuring the resistance of a grounding system are the following:

- three-pole (62% and fall of potential)
- four-pole (62% and fall of potential)
- selective
- stakeless (clamp-on)

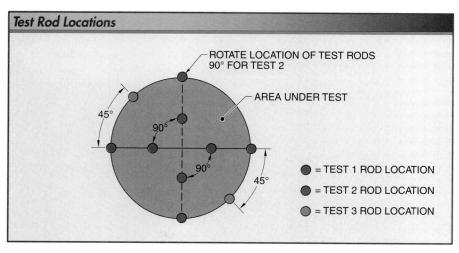

**Figure 2-4.** Test rods must be rotated 90° for the second set of readings and can be rotated at 45° to get a third set of readings as required.

Each method of measuring the resistance of an installed grounding electrode system to the earth applies Ohm's law. That is, for each method, a known or measured current and voltage are used to calculate the resistance ($R=E \div I$) grounding path to earth. Current and voltage are applied and measured differently with each method. With the three-pole, four-pole, and selective methods, stakes are driven in the ground to pass current through the ground from the grounding electrode under test as voltage measurements are taken between inner stakes. In the stakeless method, a clamp is used to induce a known voltage, and a second clamp is used to measure the current flowing within the ground system.

When using the three-pole or four-pole methods, the grounding system under test can be disconnected from the power source before any measurements can be taken. This is not a problem when testing new grounding systems that have been installed but not connected to the power source (or the power source has not been turned on). The three-pole and four-pole methods are widely accepted and have been the traditional methods of grounding system testing after initial installation of the grounding system.

If power is applied to the system, power must be removed before using the three-pole or four-pole measuring method. This requires complete knowledge of the effects disconnecting the grounding system will have when testing a system that is already powered. Using proper personal protective equipment (PPE), working with a knowledgeable partner, and advance approval are also required.

The selective and stakeless methods are newer types that do not require that system power be removed before taking the measurements. The selective method usually includes a ground tester, which includes three-pole and four-pole testing functions. This combination package allows the testing of systems before system power is applied (usually with new construction) and after system power is applied (selective method). The stakeless testing method is the easiest method to use because it involves one self-contained unit, such as a standard clamp-on ammeter, that includes both a voltage transmitter and a current meter.

**TECH TIP**

*The more earth paths used in a stakeless test, the more accurate the reading is to the actual ground resistance.*

It is important to understand the advantages, limitations, and measuring requirements of the stakeless method before attempting to use it. For example, the stakeless method must be used in systems that include multiple grounds and cannot be used to test isolated grounds that do not include a path (loop) for the applied current of the testers to flow through. However, the stakeless method works well when testing grounds that include multiple grounding paths that cannot have grounds disconnected such as utility substations and cellular signal towers.

Accurate ground measurements must be taken to ensure that an electrical system is safe for individuals and equipment. Since there is no single system that works best for all applications, it is best to learn each of the four measurement methods and their uses, advantages, and disadvantages in order to obtain the most accurate ground resistance reading.

*Ground resistance measurements can be taken on metal water pipes as required.*

## Principles of Ground Resistance Measurements

*Ohm's law* is a mathematical formula stating that the current in an electrical circuit is directly proportional to the voltage and inversely proportional to the resistance. Ohm's law is used to determine the relationship between voltage, current, and resistance in an electrical circuit. Although standard DMMs can be used with a power

supply, they are not required when a ground testing meter is used because the ground testing meter includes a built-in power supply, voltmeter, ammeter, and circuitry to take the measurements and calculate resistance.

Source voltage from a power supply is applied between the outside rods (rod 1 and rod 3). An ammeter is connected into the circuit to measure the current draw from the power supply. The current draw from the power supply is inversely proportional to the resistance of the circuit created (earth resistance). The lower the measured resistance, the higher the fault-current-carrying capacity of the circuit. Likewise, the higher the measured resistance, the lower the fault-current-carrying capacity of the circuit. **See Figure 2-5.**

A voltmeter connected between rod 1 and rod 2 measures the voltage potential difference of the earth between the two points. Rod 2 can be moved in a straight line between rod 1 and rod 3. As rod 2 is moved closer to rod 3, the voltmeter reads a higher voltage. When rod 2 is moved closer to rod 1, the voltmeter indicates a lower voltage.

Ohm's law can be used to calculate the resistance for each measurement point. Calculations are automatically performed by a ground resistance meter, which displays the resistance calculation as a measurement.

---

**TECH TIP**

*Grounding electrode conductors must be sized according to NEC® 2014 Table 250.66.*

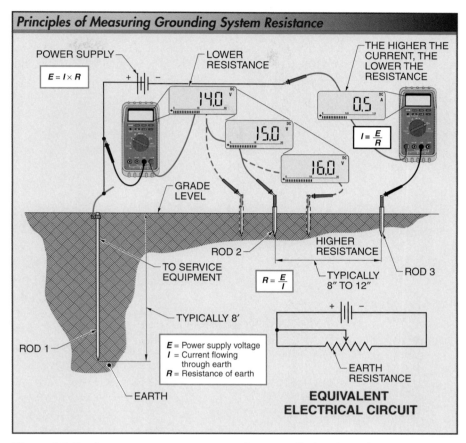

**Figure 2-5.** Resistance measurements are taken to verify that the resistance to ground of a grounding system is within required maximum resistance values set by the NEC® and can be verified by the application of Ohm's law.

For most measurements, areas are reached where the rate of increase in the resistance of the earth is low and where the resistance remains relatively constant for a set distance. The area of measurement where the resistance remains relatively constant can be referred to as the plateau area. **See Figure 2-6.**

Studies done on field tests indicate that the acceptable value of the grounding system resistance is typically when rod 2 is placed about 62% of the distance from rod 1 (grounding electrode) to rod 3. When rod 1 and rods 2 and 3 are not placed far enough apart, the measured resistance will continue to increase as rod 2 is moved closer to rod 3. There will be no leveling off (plateau) of resistance measurements. This indicates that the distance between rod 1 and rod 3 must be increased and new measurements must be taken to obtain a more accurate ground resistance measurement.

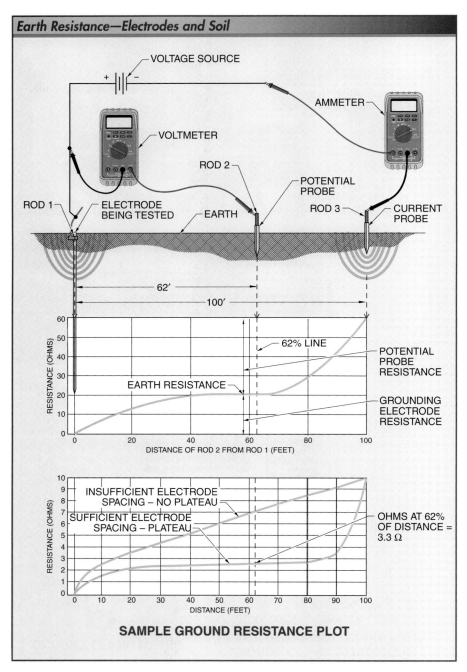

**Figure 2-6.** The area of measured resistance that remains relatively constant can be referred to as the plateau area.

## Three-Pole Ground Resistance Testing

A ground resistance meter includes a power supply, a voltmeter, an ammeter, a display for the direct readout of resistance, and all the required components for measuring earth resistance or the resistance to ground value of a grounding system. A three-pole ground resistance meter is a common meter used to test grounding systems. This method is very reliable, accurate, and can be used on any size ground system. **See Figure 2-7.**

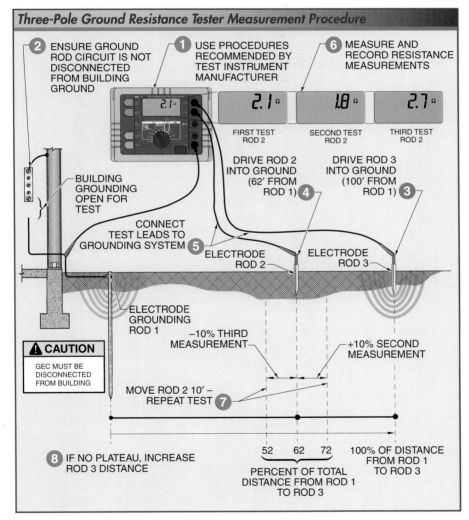

**Figure 2-7.** Before taking any earth and electrode resistance measurements, electricians must ensure that the grounding electrode is not connected to the service equipment.

There is no set distance between the placement of rod 1 and rod 3; the theoretical optimum distance will be infinite. Generally, when testing a single electrode using the three-pole earth measurement method, rod 3 is placed 100′ (62′ from rod 1) and rod 2 is driven in the earth at a distance of 31′ from rod 1 (0.62 × 50′ = 31′ for the 62% recommended point).

A 100′ spacing distance between rod 1 and rod 3 is typically adequate for most grounding systems resistance measurements except for the largest types of grounding systems.

For a large grid of electrodes consisting of several electrodes or plates that are connected, the distance between rod 1 and rod 3 must be increased to 200′ or more. Rod 2 is placed at 62% of the chosen distance.

### Three-Pole Ground Resistance Measurement Procedure

Before taking any resistance to ground measurements, the test instrument OEM operating manual must be consulted for all measuring precautions, limitations, and recommended procedures. The required personal protetcive equipment (PPE) must be worn and all safety rules must be followed as required on the construction/test site. To take three-pole ground resistance measurements, apply the following procedure:

1. Review test instrument OEM manual for recommended measuring procedures for measuring earth resistance.

2. If practical, verify that the grounding electrode and system are not connected to the building ground. **CAUTION:** If unsure of connection status, discontinue procedure until a qualified person can verify that grounding system is disconnected.

3. Drive rod 3 into the ground at distance of about 100′ from grounding electrode or grounding system being tested (rod 1).

4. Drive rod 2 into the ground at about 62′ (62%) from rod 1.

5. Connect test leads of the ground resistance tester to rods 1, 2, and 3 as indicated by the test instrument OEM.

6. Measure and record earth resistance measurement using test instrument OEM measurement procedures.

7. Move rod 2 a distance of 10′ to both sides of the 62′ point (52′ and 72′ from rod 1) and take measurements at each location. When the three readings are close, the plateau area has been determined and the 62′ reading is the reading of the grounding system resistance.

8. When the plateau for earth resistance is not determined because rod 3 is spaced too close to rod 1, increase distance between rod 1 and rod 3 and perform earth resistance test again.

### Three-Pole/Four-Pole Fall of Potential Ground Resistance Testing

Four-pole test measurements are suitable, for most applications, for measuring the resistance of a grounding system in order to meet the minimum resistance requirement listed by codes (typically 25 Ω for rod, pipe, and plate electrodes), provided the test instrument leads and connections are acceptable. Although 25 Ω may be a minimum specified value for a rod, pipe, or plate electrode, a good grounding system should have a resistance of 5 Ω or less. In addition, some applications, such as with telecommunications equipment, require that the grounding system meet a 5 Ω or less minimum specification. **See Figure 2-8.**

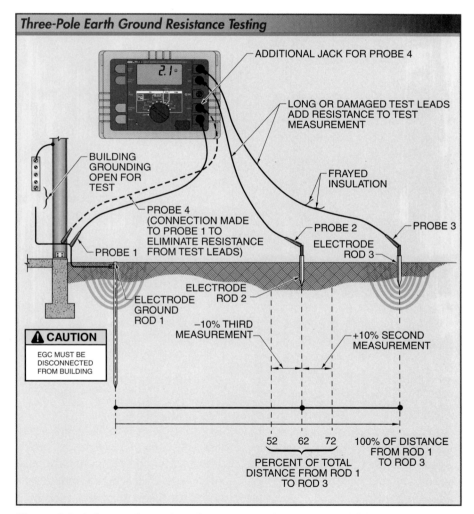

## Three-Pole Earth Ground Resistance Testing

ADDITIONAL JACK FOR PROBE 4

LONG OR DAMAGED TEST LEADS ADD RESISTANCE TO TEST MEASUREMENT

BUILDING GROUNDING OPEN FOR TEST

FRAYED INSULATION

PROBE 4 (CONNECTION MADE TO PROBE 1 TO ELIMINATE RESISTANCE FROM TEST LEADS)

PROBE 1

PROBE 2

PROBE 3

ELECTRODE ROD 3

ELECTRODE ROD 2

ELECTRODE GROUND ROD 1

**CAUTION**

EGC MUST BE DISCONNECTED FROM BUILDING

−10% THIRD MEASUREMENT

+10% SECOND MEASUREMENT

52  62  72   100% OF DISTANCE FROM ROD 1 TO ROD 3

PERCENT OF TOTAL DISTANCE FROM ROD 1 TO ROD 3

**Figure 2-8.** A three-pole earth resistance meter includes an additional jack in which another test lead is connected from the meter to the ground electrode under test to eliminate the resistance of the test leads from the displayed resistance measurement.

When a low-ground resistance measurement must be made, the resistance of the test leads must be eliminated because they add to the resistance measurement and increase the measured value. A four-pole earth resistance meter includes an additional jack in which another test lead is connected from the meter to the ground electrode under test to eliminate the resistance of the test leads from the displayed resistance measurement.

*Four-pole earth resistance meters work best for low-resistance measurements because they have an additional jack that helps eliminate resistance from the test leads in the displayed resistance measurement.*

### Selective Earth Ground Resistance Testing

The selective ground resistance testing method can be performed without removing the grounding elecrode from the building electrical grounding system. This method is used to measure individual ground electrodes of different types such as ground rods, plates, grids, and wire mesh. This method is used with grounding systems that have parallel grounds such as utility substations, transmission and distribution towers, and other commercial and industrial applications.

The selective method is similar to the three-pole method using three probes but also uses a clamp-on current transformer that eliminates the effects of parallel connected grounds from the measurement. Therefore, it measures only the electrode under test. The current transformer is placed around the earth conductor to measure the current flow to ground from the test probes through the grounding electrode under test. Once the transformer is connected, the same measuring procedures used in the three-pole method are applied.

To determine the resistance of the tower to ground, each ground point must be measured individually, and the laws of parallel-connected resistance are applied. **See Figure 2-9.** For example, if the measured resistances of the four test points (tower leg ground) are approximately 45 Ω, 36 Ω, 30 Ω, and 90 Ω, the effective resistance of the total ground system is 10.6 Ω.

### Stakeless Ground Resistance Testing

The stakeless testing method, as with the selective method, can be performed without removing the ground from the power supply. This method requires the use of two clamps: one to transmit a known voltage and the other to measure the current when a ground tester that also includes the three-pole/four-pole

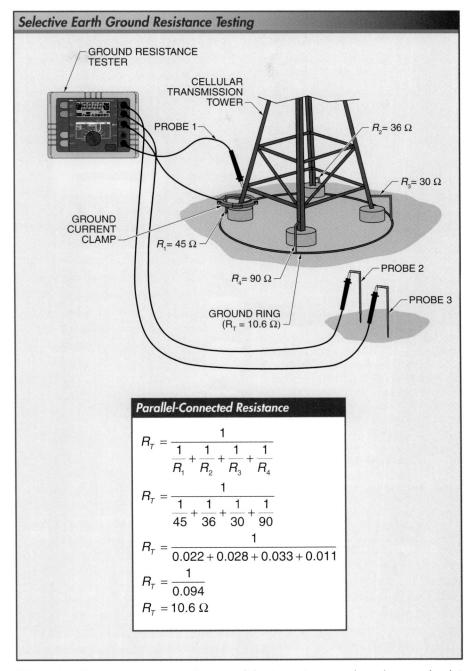

**Figure 2-9.** To determine the resistance of the tower to ground, each ground point must be measured individually and the laws of parallel-connected resistance applied.

## Finding Total Resistance Using Calculators

The total resistance of a parallel circuit containing three or more resistors is found by applying the following formula:

$$R_T = \cfrac{1}{\cfrac{1}{R_1} + \cfrac{1}{R_2} + \cfrac{1}{R_3} + \cfrac{1}{R_4}}$$

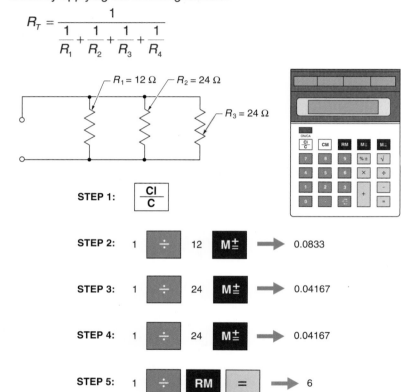

STEP 1:

STEP 2:  1  ÷  12  M± ➡ 0.0833

STEP 3:  1  ÷  24  M± ➡ 0.04167

STEP 4:  1  ÷  24  M± ➡ 0.04167

STEP 5:  1  ÷  RM  = ➡ 6

### CALCULATOR

A calculator can be used to easily apply this formula when determining the total resistance of a parallel circuit containing three or more resistors. **See Calculator.** The total resistance in a parallel circuit containing three resistors is found using a calculator by applying the following procedure:

1. Clear the calculator so that it reads "0". Ensure it does not read "0M".
2. Hit keys **1**, ÷, $R_1$ value, and **M+** to enter first resistor value into memory.
3. Hit keys **1**, ÷, $R_2$ value, and **M+** to enter second resistor value into memory.
4. Hit keys **1**, ÷, $R_3$ value, and **M+** to enter third resistor value into memory.
5. Hit keys **1**, ÷, **MR** (or **RM**), and =. The calculator displays the total resistance of the three resistors connected in parallel.

and earth resistance tester in one unit is used. Also available are test units that include both transformers in one inclusive unit. The inclusive unit cannot be used to measure earth resistance or perform the three-pole/four-pole ground test. However, the inclusive unit can be used as a current clamp to measure current, similar to a standard clamp-on ammeter, or the measurement of any leakage current flowing in the ground system. **See Figure 2-10.**

The stakeless testing method is the only type of ground testing method that does not require the use of stakes or testing probes. Therefore, it can be used in locations where using a stake is difficult or impossible such as building interiors or areas without accessible unpaved or ground (dirt) exposure. When taking the measurement, it is important to know the type of grounding electrode system (rod, building, ring, water pipe, etc.) that is part of the total building grounding system because the stakeless method measures the entire grounding system that includes all grounds, the earth, ground system bonding, and connections.

### *Stakeless Ground Resistance Testing Procedure*

Before taking a resistance measurement, it is best practice to first take a ground leakage current measurement. *Leakage current* is current that is not functional, including current in earth conductors and enclosures. Leakage current can flow through conductors and insulation. For the protection of personnel, there is a limit between 4 mA and 6 mA. Leakage current measurements can be taken when using a ground testing meter that includes a specific leakage current measurement function or by using a separate current clamp-on ammeter. When any leakage current is measured, the problem must be found and corrected.

The use of a stakeless ground tester requires complete knowledge of how and where to take measurements and the significance of the measurements. The test instrument OEM operating manuals should always be consulted for the test instrument model in use. Additional information, such as test instrument application is also usually provided on the OEM website.

To take ground resistance measurements on a grounding system with multiple parallel grounding electrodes such as transformers, utility grounds, transmission tower grounds, and communication ground systems, apply the following procedure:

1. Determine the best positions to take ground resistance measurements. **See Figure 2-11.**

2. Take current measurement in all components of the grounding system including leakage current measurement using a grounding meter or a separate current clamp-on meter. Currents over 1 A indicate a problem that must be addressed immediately. Also, ground resistance tester OEMs specify the maximum current allowed in which a meter can take an accurate measurement (typically around 5 A).

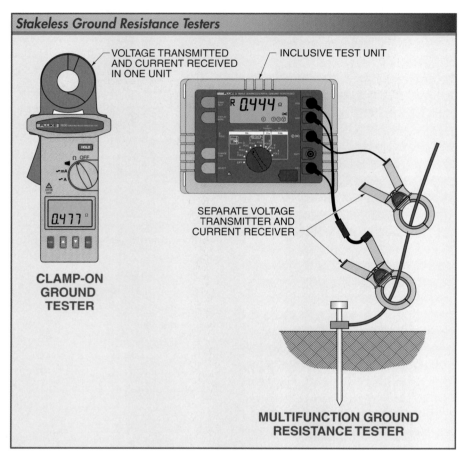

**Figure 2-10.** Stakeless earth ground resistance tests can be performed without removing the ground from the power supply.

**CAUTION:** All current measurements must be considered. Even leakage currents of a few milliamps can cause electrical shock.

3. Set meter to measure the resistance of grounding electrode conductor, place jaws of the meter clamp around the ground point under test, and record measurement.

4. Take additional measurements at each grounding point as required. For example, if there are three grounding electrodes, take a measurement at each one and at the common tie point. The measurements will be different because the tester measures the

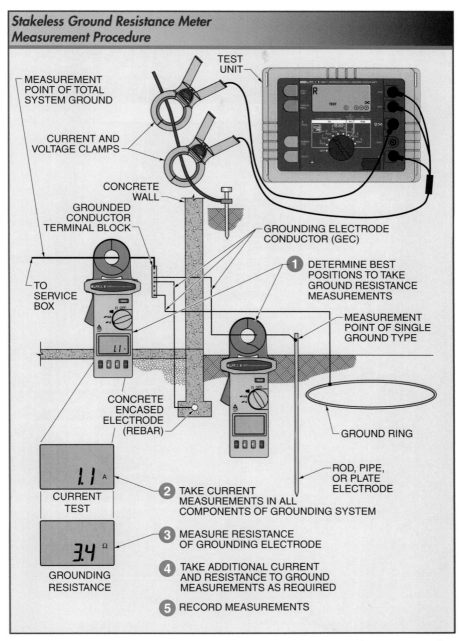

**Figure 2-11.** Ground resistance measurements are taken on conductors, grounding electrodes, and connections that include multiple grounds, such as service entrances, transformers, utility grounds, transmission tower grounds, and communication ground systems.

earth ground resistance at that point in relationship to all the grounds. The total ground at the common tie point represents the total earth ground. It is lower than the individual ground resistance measurements because all ground systems are connected in parallel. In any parallel circuit, the total resistance is always lower than any one individual resistance.

5. Record measurements and locations where taken.

Other considerations for using earth ground testers include the following:

- An earth ground tester uses batteries that must be in good operating condition. If the low battery (LO-BAT) symbol appears on the meter, the batteries must be replaced immediately. Since earth ground tests are often made in remote locations and outdoors, carrying an extra set of batteries is good practice.

- For proper measurement, use only the ground rod electrodes provided by the test instrument OEM since different material types (copper, stainless steel, aluminum, and steel) and rod size can affect total rtesistance measurements.

- Ground rod electrodes must be driven into the ground. They must never be hammered into the ground because they may bend or become damaged. Also, they must be cleaned well after each use because any dry earth stuck on them from previous usage can affect measurements.

- Ground rod electrodes and earth ground testers are connected with cables supplied by the test instrument's OEM. It is important to keep the cables in good working order. Any damage that reduces the conductor size, such as broken wire strands, increases cable resistance and affects the measurement. Cables must always be replaced with OEM-specified replacement type and size cables.

- Accessories for earth ground testers include rigid protective storage and carrying cases. Using the cases when the units are not in use helps keep the units dry, clean, and in good working condition. While protective cases help prevent damage to the test equipment, extra care should be taken to keep the meters and accessories in a safe location.

# Grounding Problems and Solutions

Proper grounding helps prevent electrical shock, fires, equipment damage, and power quality problems that can cause systems, circuits, and equipment to operate improperly. To prevent such problems, grounding systems and problems must be understood, identified, tested, corrected, and retested after any corrections or modifications. Even properly designed and installed grounding systems must be periodically retested to ensure they are still operating safely and efficiently. Testing requires knowledge of what to look for, confidence that the meter and test are best for the application, the proper meter usage for the test instrument used, and most important, an understanding of the test measurements.

## TROUBLESHOOTING

To ensure that a grounding system is properly installed and operating as designed to meet code and industry standards as well as original equipment manufacturer (OEM) and customer specifications, test instruments must be used. Test instruments must also be used when determining the cause of problems. *Troubleshooting* is the systematic diagnosis of a system to locate any fault or problem. A skilled troubleshooter follows a logical plan to find the problem quickly and efficiently. When the problem is found and corrected, the system must be retested to ensure that the problem was indeed corrected. Routine preventive maintenance and retesting are performed to prevent future problems. Troubleshooting grounding systems and ground fault problems requires an understanding of certain criteria.

• Safety Practices — Working on or around any electrical system requires an understanding of associated hazards, how to wear, use, and maintain the needed personal protective equipment (PPE), and all applicable safety rules and procedures for the given location and application.

• Technical Knowledge and Application — Grounding terminology, electrical components, facility requirements, and test instrument procedures must be understood before taking any measurements. However, having technical knowledge of how to do something does not mean one understands what results to expect and what unexpected results also mean. In addition to safety practices, understanding the application of each test instrument and its proper use is required prior to taking any measurements. It is required to know and understand where to perform tests and take measurements, how to select measurements, and interpret measurements for the proper analysis of the system and in determining any problems. Without an understanding of the difference in measured values and specified values, it cannot be made clear that the system is operating properly. Also, the type of problem cannot be identified and the proper corrective action cannot be taken.

• Solutions—Once the problem that caused the fault is identified, determining the correct solution ensures that the problem can be corrected. A system must be retested to verify that the proper solution has been applied.

• Verification and Documentation— To know whether a system is operating properly or that a problem has been corrected and to verify that all requirements are being met, test instruments must be used. Recorded data comparisons of measured values to required or recommended values serve as documentation of system operation.

• Maintenance—Since grounding systems are designed for the prevention of electrical shock, fires, and equipment problems or damage, the system must be retested at regular intervals. This includes any time a change is made to the system, new equipment is added, or a problem such as damaged components is observed. Typically, a grounding system must be inspected annually to verify and document system operation.

## SOIL RESISTANCE PROBLEMS

Soil resistivity tests are taken to determine the location, type, and size of the earth grounding system. In most areas, the ground has low enough resistivity values, measured in Ω/cm, that one type of standard grounding system, such as ground rods, rings, or plates, can be used. However, some buildings, towers, and structures are built on poor conductive soil such as bare stone or rocky areas. In such areas, grounding electrodes can be encased in a low-resistance, noncorrosive concrete to reduce the resistance between the grounding system and the earth. **See Figure 3-1.**

Although other methods can be used to reduce the soil resistance between the grounding electrode and earth, such as treating the soil with salt mixtures, only methods that are environmentally and electrically safe can be used. Salt mixtures can increase corrosion on metal parts, cause environmental damage, and dissipate over an area, which can increase the ground resistance above the minimum safe requirements. The best solution is to encase the electrode with a specially designed concrete mixture that is noncorrosive, long-lasting, and has low resistance over the life of the grounding system.

*USDA NRCS*

*Soil quality can often be reviewed by visually inspecting soil samples. Darker soil has better resistance properties.*

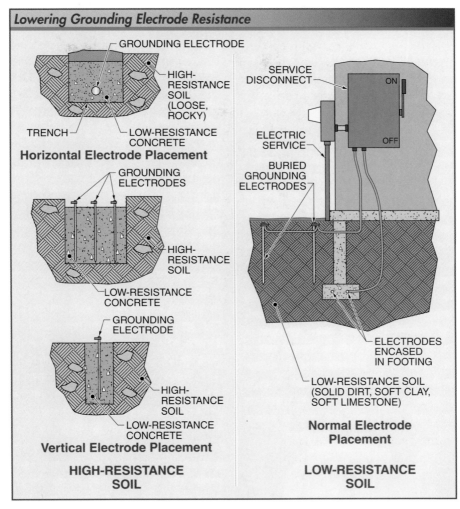

## Lowering Grounding Electrode Resistance

GROUNDING ELECTRODE

HIGH-RESISTANCE SOIL (LOOSE, ROCKY)

TRENCH

LOW-RESISTANCE CONCRETE

**Horizontal Electrode Placement**

GROUNDING ELECTRODES

HIGH-RESISTANCE SOIL

LOW-RESISTANCE CONCRETE

GROUNDING ELECTRODE

HIGH-RESISTANCE SOIL

LOW-RESISTANCE CONCRETE

**Vertical Electrode Placement**

**HIGH-RESISTANCE SOIL**

SERVICE DISCONNECT

ON

OFF

ELECTRIC SERVICE

BURIED GROUNDING ELECTRODES

ELECTRODES ENCASED IN FOOTING

LOW-RESISTANCE SOIL (SOLID DIRT, SOFT CLAY, SOFT LIMESTONE)

**Normal Electrode Placement**

**LOW-RESISTANCE SOIL**

**Figure 3-1.** When buildings, towers, and structures are built on poor conductive soil, such as bare stone or rocky areas, grounding electrodes can be encased in low-resistance, noncorrosive concrete to reduce the resistance between the grounding system and the earth.

The soil must be a low-resistance conductive part of the entire electrical system at all times. Only by testing the grounding system resistance, after installation with test instruments using the three/four-pole, selective, or stakeless method, can the true resistance of the installed grounding system be known. At times, adjustments may be required to lower system resistance greater than what is specified. Over time, experience by a technician or designer in a given area with different soil types helps determine the best type and size of grounding system to install.

## GROUNDING ELECTRODE INSTALLATION PROBLEMS

Electrical systems that are grounded shall be connected to the earth with a grounding electrode, the metal frame of a building, concrete-encased electrodes, a grounding ring, or an underground metal water pipe, per NEC® and local requirements. The grounding system is one of the most important parts of an electrical system because it can prevent electrical shock, fires, and equipment damage yet it is often overlooked once installed and not inspected or tested until a problem occurs. Grounding systems that once met the minimum resistance requirements may no longer meet those requirements. Components of a grounding system may have been damaged to the point that they are no longer part of the system. To prevent ground-related problems, the grounding system must be installed correctly and must be inspected on a regularly scheduled basis to ensure that it is functioning as designed.

*Grounding electrodes can be driven into hard, rocky surfaces with heavy-duty hammer drills.*

### Grounding Electrode System Installation

Rod, pipe, and plate electrodes shall meet NEC® requirements. Rod-type grounding equipment of stainless steel, copper, and zinc-coated steel shall be copper clad and have a ⅝″minimum diameter and length of 8′ driven vertically into noncorrosive soil with good conductivity. The top of the electrode must be at or below ground level unless protected from physical damage. Also, there must be at least 8′ of the rod and pipe electrodes in contact with the soil.

Less than optimal soil conditions may require more complex grounding systems or artificial soils to reduce the resistance to ground to required levels. The grounding system resistance to ground for a rod, pipe, or plate electrode must be 25 Ω or less. **See Figure 3-2.**

If rocky conditions prevent vertical installation, the rod electrode can be driven at an angle not to exceed 45° from vertical. Alternatively, the rod electrode may be buried in a trench that is at least 2½′ deep. Additional electrodes can be connected in parallel to reduce total resistance. If rod electrodes cannot be used or do not meet minimum requirements, alternative grounding methods such as metal building frames, a ground ring, or a grounding plate must be used. In some applications, installing a ground rod in low-resistance concrete designed for grounding systems lowers the resistance to meet code requirements. Any method used must be tested using a ground resistance meter to verify that code requirements are met.

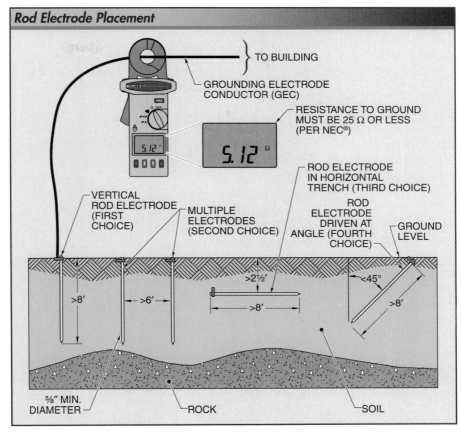

**Rod Electrode Placement**

**Figure 3-2.** A grounding rod electrode with a minimum diameter of ⅝″ and length of 8′ is placed into noncorrosive soil with the top of the electrode at or below ground level. There must be at least 8′ of the electrode in contact with the soil.

## Using Multiple Grounding Electrodes

Per the NEC® if one rod, pipe, or plate electrode exceeds the 25 Ω limit of resistance to ground, additional electrodes can be added to the system to lower the total resistance. Afterward, the resistance to ground must be measured. Resistance is lowered by approximate percentages as each additional rod with the same individual resistance is added. The second rod lowers the total resistance to approximately 60% of the first rod. The third rod lowers the total resistance to approximately 40% of the first rod. The fourth rod lowers the total resistance to approximately 33% of the first rod. Multiple electrodes must be at least 6′ apart and connected together at the top. **See Figure 3-3.**

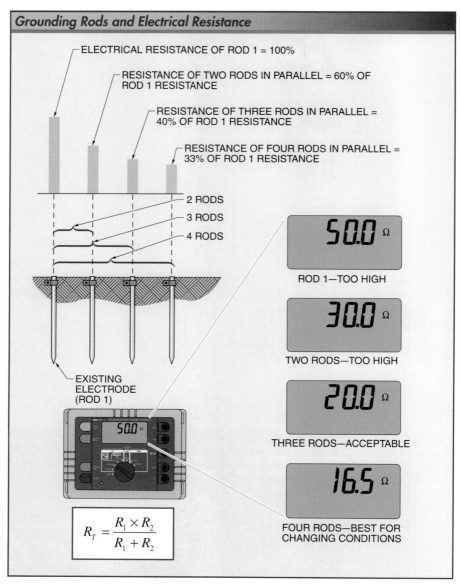

**Figure 3-3.** If one electrode exceeds the 25 Ω limit of resistance to ground, additional electrodes can be added to the system to lower the total resistance.

*Note*: The result of a grounding system (rod, ring, plate, etc.) is to ensure that the minimum resistance (usually 25 Ω) is met. In addition to adding more rods, their size or capacity may be increased as required, as long as they are all interconnected (bonded) per code requirements.

## OPERATIONAL PROBLEMS

A properly designed and installed grounding system must operate properly. In most cases, the grounding system does not operate properly. Unfortunately, a grounding system may not show any signs of a problem until there is an electrical shock, equipment is damaged, or equipment does not operate properly. It is important to understand how the grounding system must operate, what faults may occur, and how to test and correct a fault.

### Ground-Loop Problems

A grounding electrode system is installed at the main electrical service or at the source of a separately derived system. A *separately derived system (SDS)* is an electrical system that supplies electrical power derived or taken from transformers, storage batteries, photovoltaic systems, wind turbines, or generators. The vast majority of SDSs are produced by the secondary side of a power distribution transformer.

An SDS is typically used to establish a new voltage level, lower the power source impedance, or isolate parts of the power distribution system. Because an SDS does not have direct electrical connections to any other part of a supply distribution system (transformers magnetically couple), a new ground reference is required. A proper ground reference is required for safety and proper equipment operation and is established by making a ground connection between the SDS and ground.

*Grounding system conductors are typically identified by the presence of green-colored insulation.*

The earth grounding is established through a grounding electrode such as a rod or steel building frame that has been effectively grounded. The neutral to ground connection must be made at the transformer or the main service panel only. The neutral to ground connection is made by connecting the neutral bus to the ground bus with a main bonding jumper and to the grounding electrode system by a grounding electrode conductor. **See Figure 3-4.** A *main bonding jumper (MBJ)* is a connection in a service panel that connects the equipment grounding conductor (EGC), the grounding electrode conductor (GEC), and the grounded conductor (neutral conductor).

An *equipment grounding conductor (EGC)* is an electrical conductor that provides a low-impedance grounding path between electrical equipment and enclosures within a distribution system. A GEC connects grounded parts of a power distribution system (equipment grounding conductors, ground conductors, and all metal parts) to the NEC®-approved grounding system.

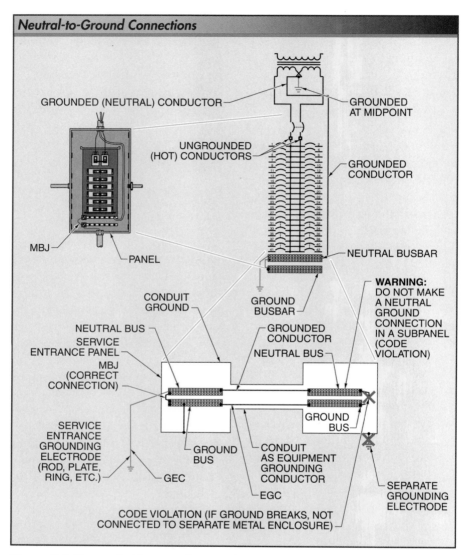

**Figure 3-4.** The neutral-to-ground connection is made by connecting the neutral bus to the ground bus with an MBJ and to a grounding electrode.

Neutral-to-ground connections must not be made in any subpanels, receptacles, or equipment. When a neutral-to-ground connection is made anywhere except in the main service panel, a parallel flow path for the normal return current from a load is created. The parallel flow path allows current to flow through metal parts of the system. The NEC® does not allow neutral-to-ground connections that create ground loops because it can cause electrical shock and power quality problems.

In addition to not making neutral-to-ground connections in subpanels, no additional earth grounds such as grounding electrodes can be established. An additional, separate, isolated grounding electrode creates two ground references that are typically at different voltage potentials.

A *ground loop* is an electrical circuit that has more than one grounding point connected to earth ground, with a voltage potential difference between the grounding points high enough to produce a circulating current in the grounding system. The two grounding electrodes result in current circulating and forming a ground loop between the two grounding electrodes in an attempt to equalize the difference in voltage potential. Current circulation is caused by current that flows from a higher voltage potential to a lower voltage potential. A voltage potential exists because there is a difference in impedance (total resistance, inductance, and capacitance) between the two ground points. **See Figure 3-5.**

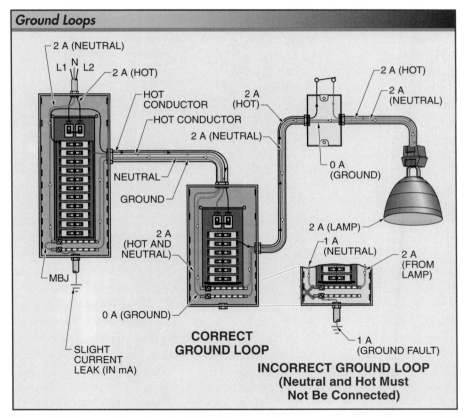

Figure 3-5. A ground loop is a circuit that has more than one grounding point connected to earth ground, with a voltage potential difference between the grounding points high enough to produce a circulating current in the grounding system.

### Circuit Ground-Fault Current

All electrical circuits are designed with a normal path for current to flow. For example, the normal path of current flow for a standard 115 VAC circuit is from a hot conductor, through the load, and back through the neutral conductor. Current should never flow through branch circuit ground conductors and/or metal at any time. Current must only flow through the ground conductor when a fault occurs in the circuit. Typically, faults that cause current to flow through a ground conductor include short circuits, insulation breakdown, moisture, corrosion, damaged wires, and illegal neutral to ground connections.

Current that does not take a designed path is referred to as leakage current. Leakage current is current that leaves the normal path of current flow (hot to neutral) and flows through a ground path. Leakage current can be in equipment, branch circuits, or any place where current flows to the grounding system due to a fault. For example, if there is insulation breakdown between a hot and ground conductor or grounded metal, current can flow from the hot conductor to the ground conductor.

The amount of current flow can be small (in μA or mA) or large (up to several amps). If current is high enough, the fuse opens or the circuit breaker trips. It is the smaller amount of leakage current that causes problems such as electrical shock because it can go undetected until it increases to the point of tripping the circuit breaker. **See Figure 3-6.**

Theoretically, there should not be any leakage current in equipment or branch circuits. Some leakage current to ground will be present through the building grounding system. Some leakage current exists because the utility power source is grounded at the transformer and the building grounding system is grounded at the service entrance. The earth between the two grounding points can allow some current to flow through the ground.

Since the points at which the utility is grounded and the service is grounded with a low resistance conductor, the conductor carries most of the system neutral current back to the transformer. The amount of current flow through the ground electrode back to the transformer is typically in the range of 5 mA to 100 mA. Any current over 5 mA should indicate the source of leakage current. A grounding system should be tested inside the power panel. This is to inspect the system wiring and branch current on each hot, neutral, and ground to understand system operation. High current can cause problems, and high ground leakage current is typically caused by poor, loose, or damaged neutral connections that increase the total resistance of the neutral conductor. **See Figure 3-7.**

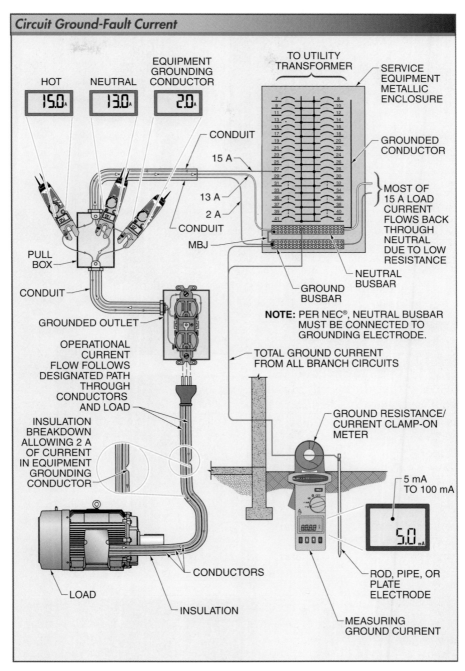

## Circuit Ground-Fault Current

**HOT** **15.0** A

**NEUTRAL** **13.0** A

**EQUIPMENT GROUNDING CONDUCTOR** **2.0** A

CONDUIT

15 A

13 A

2 A

CONDUIT

MBJ

PULL BOX

CONDUIT

GROUNDED OUTLET

OPERATIONAL CURRENT FLOW FOLLOWS DESIGNATED PATH THROUGH CONDUCTORS AND LOAD

INSULATION BREAKDOWN ALLOWING 2 A OF CURRENT IN EQUIPMENT GROUNDING CONDUCTOR

LOAD

CONDUCTORS

INSULATION

TO UTILITY TRANSFORMER

SERVICE EQUIPMENT METALLIC ENCLOSURE

GROUNDED CONDUCTOR

MOST OF 15 A LOAD CURRENT FLOWS BACK THROUGH NEUTRAL DUE TO LOW RESISTANCE

NEUTRAL BUSBAR

GROUND BUSBAR

**NOTE:** PER NEC®, NEUTRAL BUSBAR MUST BE CONNECTED TO GROUNDING ELECTRODE.

TOTAL GROUND CURRENT FROM ALL BRANCH CIRCUITS

GROUND RESISTANCE/ CURRENT CLAMP-ON METER

5 mA TO 100 mA

**5.0** mA

ROD, PIPE, OR PLATE ELECTRODE

MEASURING GROUND CURRENT

**Figure 3-6.** Ground-fault current can be tested with test instruments such as ammeters and ground resistance/current meters.

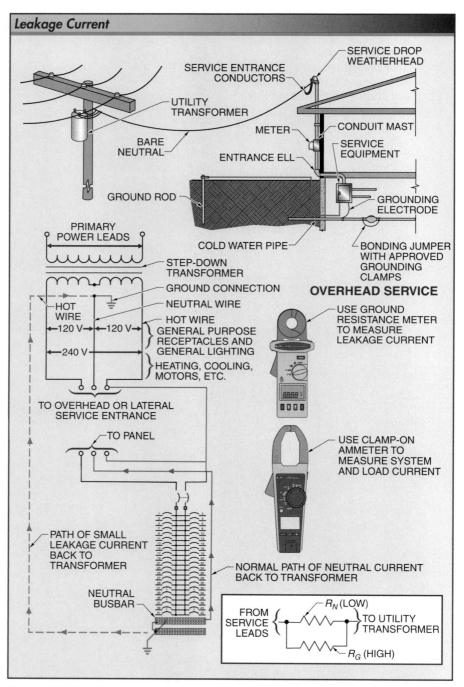

**Figure 3-7.** Electrical current flows to a difference potential between two points.

# System Troubleshooting and Preventive Maintenance

Grounding systems and components are part of an entire electrical system from generation to each final point of delivery and usage. This includes the utility grounding system, ground electrodes, lightning grounds, building grounds, equipment grounds, and circuit grounds. A fault such as an open circuit, short circuit, high-resistance connection, or improper bond can cause problems within the system and produce an unsafe and dangerous condition. Grounding system troubleshooting and preventive maintenance requires using several different test instruments in addition to an earth ground tester to locate any potential problems or faults.

## LOOSE CONNECTIONS

When current flows through any device that has resistance, there is a voltage drop across that device, and heat is produced because of the power produced ($E = I \times R$). Since all conductors (wires), splices, and connections have some resistance, there is always some power/heat produced within the device. Properly sized conductors and properly made connections have low resistance and thus produce little heat. Conductors that are undersized and connections that are loose have higher resistance and thus produce much heat. Even conductors that are properly sized produce heat because they have resistance. The amount of heat produced depends on the amount of current flow through the conductors and connections. The higher the amount of current flow through the conductor or connection, the more heat is produced in the conductors/connections.

A *thermal imager* is a device that detects heat patterns in the infrared-wavelength spectrum without making direct contact with equipment. A thermal imager can be used to observe or measure heat produced by conductors and connections. Heat patterns are indicated by the colors blue (coolest), green, yellow, orange, and red (hottest). The temperature can also be displayed on the thermal imager screen.

All conductors/connections that have current flowing through them emit heat. As long as the current is within the designed limits, the heat produced cannot cause a problem. Since little (a few mA) or no current should flow through a grounded conductor, no heat should be produced or observed in the conductor or connection. When there is any indication of heat, there is a problem that must be investigated.

### TECH TIP

*National Electrical Code® (NEC®) 2014 Article 250, parts III, IV, V, and VI cover grounding conductor connections.*

Measuring voltage drop across a connection with a voltmeter or digital multimeter (DMM) determines a good or poor connection. The higher the measured voltage drop across a connection, the higher the resistance of the connections and the lower the quality of the splice. Measuring the ground system resistance (conductors and connections) with a clamp-on ground tester indicates a problem within the system. **See Figure 4-1.** If a problem, such as high resistance, is measured, individual parts of the ground system must be inspected and tested. The problem is usually a poor connection, so all connections should be inspected. Other problems can be that the resistance of the ground electrodes has increased due to dry soil or corrosion on the rods.

*Thermal imagers can be used to safely detect unseen problems with electrical systems and equipment without making contact.*

## IMPROPER GROUND CONNECTIONS

The earth ground is connected at the main service equipment or at the source of a separately derived system (SDS). An SDS supplies electrical power derived or taken from transformers, storage batteries, photovoltaic systems, wind turbine generators, or other generators. **See Figure 4-2.**

Because an SDS does not have electrical connections to any other part of the distribution system, a new ground reference is required. The new ground reference must be connected back to the main building ground electrode and not to a new ground electrode. Connecting back to the main building ground electrode bonds the grounding system to one common earth grounding point (electrode).

If the output of an SDS is not grounded, the system loads continue to work. However, a dangerous condition exists. A voltmeter is used to verify that the SDS output is grounded as required. **See Figure 4-3.** A voltmeter reads the output of the SDS regardless of whether the system has been grounded.

For example, to test for a grounded secondary, the voltmeter must be connected between the two power lines to verify the SDS voltage output. The voltmeter should display the output voltage regardless of whether the SDS secondary is grounded. If one of the power lines has been grounded, the voltmeter will measure the SDS secondary output between the hot (fused) conductor and ground. If one of the power lines is not grounded, the meter cannot measure a fixed voltage between any power output line and ground. The reading varies because the meter is only reading ghost voltage. A *ghost voltage reading* is a reading on a voltmeter that is not connected to an energized circuit. Ghost voltage can also be displayed if only one meter lead is connected to an energized circuit and the other meter is not connected to any point that is energized or grounded.

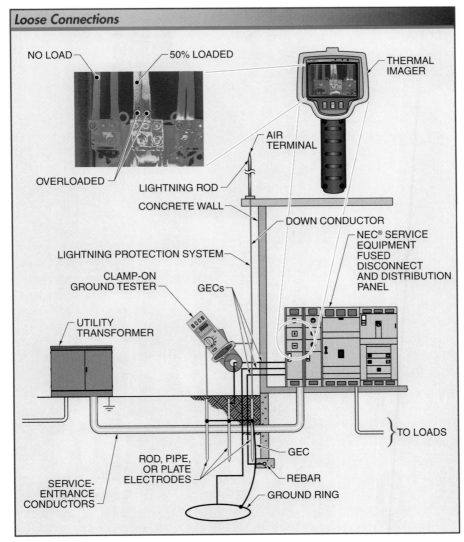

**Figure 4-1.** Loose connection problems within a grounding system can be located and measured with clamp-on ground testers and thermal imagers.

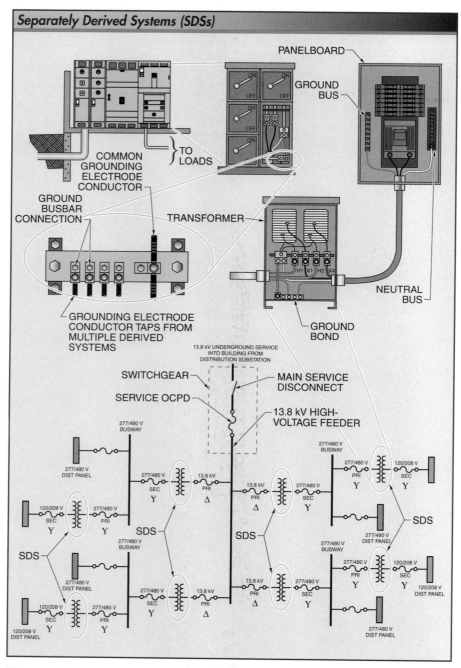

**Figure 4-2.** The grounding electrode is connected at the service equipment and at an SDS.

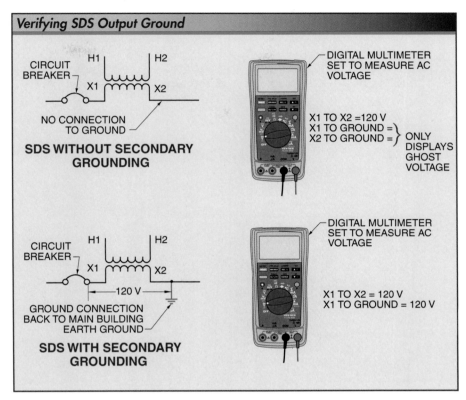

**Figure 4-3.** A voltmeter or similar test instrument is used to verify that the SDS output is grounded as required.

## MULTIPLE NEUTRAL-TO-GROUND CONNECTIONS

Neutral-to-ground connections must not be made in any subpanels, receptacles, or equipment. If a neutral-to-ground connection is made, a parallel path for the normal return current from system loads is created. The parallel path allows current to flow through metal parts of the system thus creating a dangerous condition. All grounding points must be connected back to the main earth ground electrode. **See Figure 4-4.**

Measuring the amount of current on a ground system can be helpful when inspecting building electrical systems or troubleshooting the system to locate a fault. The ground current is measured using a clamp-on ground tester set to measure current. Clamp-on ground testers are designed to measure small amounts of ground current and must not be used to measure load, branch circuit, or power conductor current. A clamp-on ammeter is used to measure load, branch circuit, or power conductor current. The measured ground current is the highest at the ground electrode because it will be the sum of all ground currents.

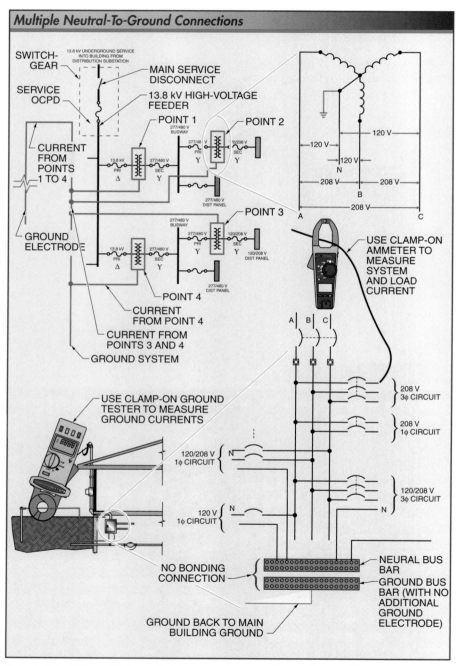

**Figure 4-4.** With multiple neutral-to-ground connections, all grounding points must be connected back to the main earth ground electrode.

## HIGH GROUND CURRENT— INSULATION BREAKDOWN

Electrical conductors are covered with insulating material. Insulation prevents current from flowing outside its designated path through the conductor to exposed grounded or ungrounded metal parts or a grounding system. Insulation must have high enough resistance to prevent current from flowing through it and causing electrical shock or a fire, tripping a circuit breaker, or blowing a fuse. A small amount of leakage current flows through most insulation. Leakage current increases as the insulation breaks down because of moisture, extreme temperatures, oil, vibration, pollutants, and mechanical stress or damage.

As insulation resistance decreases, leakage current to ground increases. Before leakage current even gets high enough to open a circuit breaker or fuse, it can cause electrical shock or a spark that could cause a fire. Electrical test instruments are used to test variables in electrical systems, loads, insulation, and grounds. **See Figure 4-5.** Electrical test instruments include the following:

- A megohmmeter (insulation tester) is used to test the condition of insulation with the power off. A megohmmeter is connected to ground and each conductor that is designed to carry current to the load.

- A clamp-on ground tester is used to measure leakage current (low current setting) and to measure current through the grounding system with the power on. Leakage current increases as measurements are taken from individual circuits and loads back to the building main ground electrode.

- A clamp-on ammeter is used to measure the amount of current draw of individual loads, individual branch circuits, or main power line feeds with the power on.

- A DMM, clamp-on ammeter, or any meter that measures resistance is used to measure the resistance of individual components or loads with the power off.

Testing or troubleshooting any electrical system requires taking several different types of measurements to completely understand how the system, circuits, and loads are operating. An individual test may identify a problem or faulty component, but it may not identify other problems that can or are causing additional problems. When testing or troubleshooting an electrical system, the circuit or load voltage, current, and grounding system resistance must always be taken to provide a starting point. In addition, other measurements may be taken to help identify other problems or to provide more information about the system. Measurements taken include the following:

- voltage, which reveals if power is present and at what level

- current, which reveals how much load is on a circuit

- ground resistance, which ensures that the ground system meets minimum resistance requirements

- ground leakage current, which identifies any potential electrical shock or fire potential problems

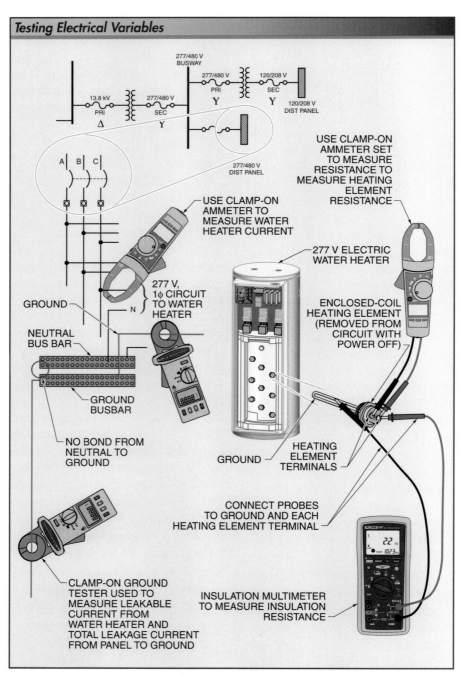

**Figure 4-5.** Electrical test instruments are used to test variables in electrical systems, loads, insulation, and grounds.

- low resistance, which identifies the level of resistance of splices, loads, etc. (*Note*: Use standard DMM set to measure resistance.)
- high resistance, which reveals the condition of insulation *Note*: Use megohmmeter or insulation resistance tester.
- power, which determines the cost of operation (W), size of transformers (VA), and efficiency (PF)

*An insulation resistance tester can be used to take insulation resistance readings on electrical equipment and systems.*

## MEASURING GROUND SYSTEM CURRENT

Some faults in an electrical system are visible, such as a burned out lamp, some faults require testing, such as a circuit breaker that keeps tripping, and others may require multiple tests with different meters at different locations. Ground system testing and trouble-shooting requires that measurements

be taken at several locations including hot, neutral, and ground conductors to understand how the system is operating and whether a problem exists. Tests include taking voltage and current measurements at the load, panels, and switchgear. **See Figure 4-6.**

A set of wireless meters, such as wireless DMMs, can be used to measure and monitor multiple readings from one central location. A wireless DMM can be used that takes readings from remote meters and displays their measurements on one meter. The meter can display meter measurements it is taking as well as display three other wireless remote modules located throughout the system. Remote meters can measure and transmit voltage, current, and temperature measurements to one meter, which allow systems to be monitored at several locations. They also provide additional safety in that the operator at the main meter can monitor readings that are already connected to a hazardous location.

### TECH TIP

*A wireless DMM displays meter measurements as well as readings from up to three wireless modules from distances as far as 20 m (66').*

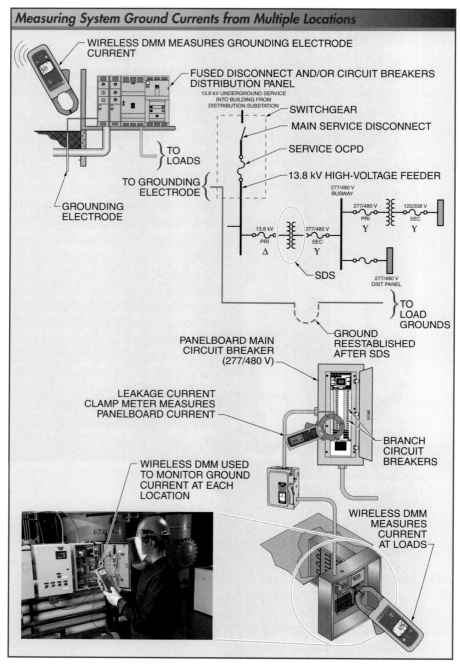

**Measuring System Ground Currents from Multiple Locations**

WIRELESS DMM MEASURES GROUNDING ELECTRODE CURRENT

FUSED DISCONNECT AND/OR CIRCUIT BREAKERS DISTRIBUTION PANEL

13.8 kV UNDERGROUND SERVICE INTO BUILDING FROM DISTRIBUTION SUBSTATION

SWITCHGEAR

MAIN SERVICE DISCONNECT

TO LOADS

SERVICE OCPD

TO GROUNDING ELECTRODE

13.8 kV HIGH-VOLTAGE FEEDER

277/480 V BUSWAY

GROUNDING ELECTRODE

277/480 V PRI

120/208 V SEC

Y

Y

13.8 kV PRI

Δ

277/480 V SEC

Y

Y

SDS

277/480 V DIST PANEL

TO LOAD GROUNDS

GROUND REESTABLISHED AFTER SDS

PANELBOARD MAIN CIRCUIT BREAKER (277/480 V)

LEAKAGE CURRENT CLAMP METER MEASURES PANELBOARD CURRENT

BRANCH CIRCUIT BREAKERS

WIRELESS DMM USED TO MONITOR GROUND CURRENT AT EACH LOCATION

WIRELESS DMM MEASURES CURRENT AT LOADS

**Figure 4-6.** A set of wireless meters, such as wireless DMMs and clamp meters, can be used to measure and monitor multiple readings from one central location.

## Power Formulas —1φ, 3φ

| Phase | To Find | Use Formula | Example | | |
|-------|---------|-------------|---------|------|----------|
| | | | Given | Find | Solution |
| 1φ | I | $I = \dfrac{VA}{V}$ | 32,000 VA, 240 V | I | $I = \dfrac{VA}{V}$ $I = \dfrac{32,000\ VA}{240\ V}$ $I = \textbf{133 A}$ |
| 1φ | VA | $VA = I \times V$ | 100 A, 240 V | VA | $VA = I \times V$ $VA = 100\ A \times 240V$ $VA = \textbf{24,000 VA}$ |
| 1φ | V | $V = \dfrac{VA}{I}$ | 42,000 VA 350 A | V | $V = \dfrac{VA}{I}$ $V = \dfrac{42,000\ VA}{350\ A}$ $V = \textbf{120 V}$ |
| 3φ | I | $I = \dfrac{VA}{V \times \sqrt{3}}$ | 72,000 VA, 208 V | I | $I = \dfrac{VA}{V \times \sqrt{3}}$ $I = \dfrac{72,000\ VA}{360\ V}$ $I = \textbf{200 V}$ |
| 3φ | VA | $VA = I \times V \times \sqrt{3}$ | 2 A, 240 V | VA | $VA = I \times V \times \sqrt{3}$ $VA = 2 \times 416$ $VA = \textbf{832 VA}$ |

## Three-Phase Voltage Values

| |
|---|
| For 208 V × 1.732, use 360 |
| For 230 V × 1.732, use 398 |
| For 240 V × 1.732, use 416 |
| For 440 V × 1.732, use 762 |
| For 460 V × 1.732, use 797 |
| For 480 V × 1.732, use 831 |
| For 2400 V × 1.732, use 4157 |
| For 4160 V × 1.732, use 7205 |

## Power Formula Abbreviations and Symbols

| | |
|---|---|
| P = Watts | V = Volts |
| I = Amps | VA = Volt Amps |
| A = Amps | φ = Phase |
| R = Ohms | √ = Square Root |
| E = Volts | |

## Ohm's Law and Power Formula

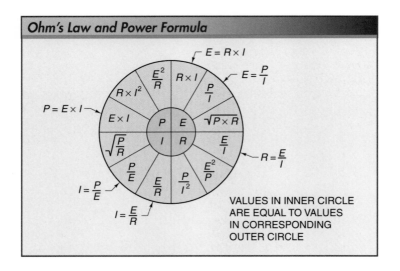

$$E = R \times I$$
$$E = \frac{P}{I}$$
$$P = E \times I$$
$$I = \frac{P}{E}$$
$$I = \frac{E}{R}$$
$$R = \frac{E}{I}$$

VALUES IN INNER CIRCLE
ARE EQUAL TO VALUES
IN CORRESPONDING
OUTER CIRCLE

## Voltage Conversions

| To Convert | To | Multiply By |
|---|---|---|
| rms | Average | 0.9 |
| rms | Peak | 1.414 |
| Average | rms | 1.111 |
| Average | Peak | 1.567 |
| Peak | rms | 0.707 |
| Peak | Average | 0.637 |
| Peak | Peak-to-peak | 2 |

## Typical Motor Efficiencies

| HP | Standard Motor (%) | Energy-Efficient Motor (%) | HP | Standard Motor (%) | Energy-Efficient Motor (%) |
|---|---|---|---|---|---|
| 1 | 76.5 | 84.0 | 30 | 88.1 | 93.1 |
| 1.5 | 78.5 | 85.5 | 40 | 89.3 | 93.6 |
| 2 | 79.9 | 86.5 | 50 | 90.4 | 93.7 |
| 3 | 80.8 | 88.5 | 75 | 90.8 | 95.0 |
| 5 | 83.1 | 88.6 | 100 | 91.6 | 95.4 |
| 7.5 | 83.8 | 90.2 | 125 | 91.8 | 95.8 |
| 10 | 85.0 | 90.3 | 150 | 92.3 | 96.0 |
| 15 | 86.5 | 91.7 | 200 | 93.3 | 96.1 |
| 20 | 87.5 | 92.4 | 250 | 93.6 | 96.2 |
| 25 | 88.0 | 93.0 | 300 | 93.8 | 96.5 |

## Metric Prefixes

| Multiples and Submultiples | Prefixes | Symbols | Meaning |
|---|---|---|---|
| $1{,}000{,}000{,}000{,}000 = 10^{12}$ | tera | T | trillion |
| $1{,}000{,}000{,}000 = 10^{9}$ | giga | G | billion |
| $1{,}000{,}000 = 10^{6}$ | mega | M | million |
| $1000 = 10^{3}$ | kilo | k | thousand |
| $100 = 10^{2}$ | hecto | h | hundred |
| $10 = 10^{1}$ | deka | d | ten |
| Unit $1 = 10^{0}$ | | | |
| $0.1 = 10^{-1}$ | deci | d | tenth |
| $0.01 = 10^{-2}$ | centi | c | hundredth |
| $0.001 = 10^{-3}$ | milli | m | thousandth |
| $0.000001 = 10^{-6}$ | micro | $\mu$ | millionth |
| $0.000000001 = 10^{-9}$ | nano | n | billionth |
| $0.000000000001 = 10^{-12}$ | pico | p | trillionth |

## Metric Conversions

| Initial Units | Final Units | | | | | | | | | | | |
|---|---|---|---|---|---|---|---|---|---|---|---|---|
| | giga | mega | kilo | hecto | deka | base unit | deci | centi | milli | micro | nano | pico |
| giga | | 3R | 6R | 7R | 8R | 9R | 10R | 11R | 12R | 15R | 18R | 21R |
| mega | 3L | | 3R | 4R | 5R | 6R | 7R | 8R | 9R | 12R | 15R | 18R |
| kilo | 6 | 3L | | 1R | 2R | 3R | 4R | 5R | 6R | 9R | 12R | 15R |
| hecto | 7L | 4L | 1L | | 1R | 2R | 3R | 4R | 5R | 8R | 11R | 14R |
| deka | 8L | 5L | 2L | 1L | | 1R | 2R | 3R | 4R | 7R | 10R | 13R |
| base unit | 9L | 6L | 3L | 2L | 1L | | 1R | 2R | 3R | 6R | 9R | 12R |
| deci | 10L | 7L | 4L | 3L | 2L | 1L | | 1R | 2R | 5R | 8R | 11R |
| centi | 11L | 8L | 5L | 4L | 3L | 2L | 1L | | 1R | 4R | 7R | 10R |
| milli | 12L | 9L | 6L | 5L | 4L | 3L | 2L | 1L | | 3R | 6R | 9R |
| micro | 15L | 12L | 9L | 8L | 7L | 6L | 5L | 4L | 3L | | 3R | 6R |
| nano | 18L | 15L | 12L | 11L | 10L | 9L | 8L | 7L | 6L | 3L | | 3R |
| pico | 21L | 18L | 15L | 14L | 13L | 12L | 11L | 10L | 9L | 6L | 3L | |

R = move decimal point to the right
L = move decimal point to the left

## Common Prefixes

| Symbol | Prefix | Equivalent |
|--------|--------|------------|
| G | giga | 1,000,000,000 |
| M | mega | 1,000,000 |
| k | kilo | 1000 |
| base unit | – | 1 |
| m | milli | 0.001 |
| μ | micro | 0.000001 |
| n | nano | 0.000000001 |
| p | pico | 0.00000000001 |
| Z | impedance | ohms – Ω |

## Voltage Drop Formulas – 1φ, 3φ

| Phase | To Find | Use Formula | Example |  |  |
|-------|---------|-------------|---------|------|----------|
|       |         |             | Given | Find | Solution |
| 1φ | VD | $VD = \dfrac{2 \times R \times L \times I}{1000}$ | 240 V, 40 A, 60 L, .764 R | VD | $VD = \dfrac{2 \times R \times L \times I}{1000}$ $VD = \dfrac{2 \times .764 \times 60 \times 40}{1000}$ $VD = \textbf{3.67 V}$ |
| 3φ | VD | $VD = \dfrac{2 \times R \times L \times I}{1000} \times .866$ | 208 V, 110 A, 75 L, .194 R, .866 multiplier | VD | $VD = \dfrac{2 \times R \times L \times I}{1000} \times .866$ $VD = \dfrac{2 \times .194 \times 75 \times 110}{1000} \times .866$ $VD = \textbf{2.77 V}$ |

| Raceways | |
|---|---|
| EMT | Electrical Metallic Tubing |
| ENT | Electrical Nonmetallic Tubing |
| FMC | Flexible Metal Conduit |
| FMT | Flexible Metallic Tubing |
| IMC | Intermediate Metal Conduit |
| LFMC | Liquidtight Flexible Metal Conduit |
| LFNC | Liquidtight Flexible Nonmetallic Conduit |
| RMC | Rigid Metal Conduit |
| RNC | Rigid Nonmetallic Conduit |

| Cables | |
|---|---|
| AC | Armored Cable |
| BX | Tradename for AC |
| FCC | Flat Conductor Cable |
| IGS | Integrated Gas Spacer Cable |
| MC | Metal-Clad Cable |
| MI | Mineral Insulated, Metal Sheathed Cable |
| MV | Medium Voltage |
| NM | Nonmetallic-Sheathed Cable (dry) |
| NMC | Nonmetallic-Sheathed Cable (dry or damp) |
| NMS | Nometallic-Sheathed Cable (dry) |
| SE | Service-Entrance Cable |
| TC | Tray Cable |
| UF | Underground Feeder Cable |
| USE | Underground Service-Entrance Cable |

## AC/DC Formulas

| To Find | DC | AC | | |
|---|---|---|---|---|
| | | 1φ, 115 or 220 V | 1φ, 208, 230, or 240 V | 3φ– All Voltages |
| I, HP known | $\dfrac{HP \times 746}{E \times Eff}$ | $\dfrac{HP \times 746}{E \times Eff \times PF}$ | $\dfrac{HP \times 746}{E \times Eff \times PF}$ | $\dfrac{HP \times 746}{1.73 \times E \times Eff \times PF}$ |
| I, kW known | $\dfrac{kW \times 1000}{E}$ | $\dfrac{kW \times 1000}{E \times PF}$ | $\dfrac{kW \times 1000}{E \times PF}$ | $\dfrac{kW \times 1000}{1.73 \times E \times PF}$ |
| I, kVA known | | $\dfrac{kVA \times 1000}{E}$ | $\dfrac{kVA \times 1000}{E}$ | $\dfrac{kVA \times 1000}{1.763 \times E}$ |
| kW | $\dfrac{I \times E}{1000}$ | $\dfrac{I \times E \times PF}{1000}$ | $\dfrac{I \times E \times PF}{1000}$ | $\dfrac{I \times E \times 1.73 \times PF}{1000}$ |
| kVA | | $\dfrac{I \times E}{1000}$ | $\dfrac{I \times E}{1000}$ | $\dfrac{I \times E \times 1.73}{1000}$ |
| HP (output) | $\dfrac{I \times E \times Eff}{746}$ | $\dfrac{I \times E \times Eff \times PF}{746}$ | $\dfrac{I \times E \times Eff \times PF}{746}$ | $\dfrac{I \times E \times 1.73 \times Eff \times PF}{746}$ |

$E_{ff}$ = efficiency

## Fuses and ITCBs

| Increase | Standard Ampere Ratings |
|----------|-------------------------|
| 5 | 15, 20, 25, 30, 35, 40, 45 |
| 10 | 50, 60, 70, 80, 90, 100, 110 |
| 25 | 125, 150, 175, 200, 225 |
| 50 | 250, 300, 350, 400, 450 |
| 100 | 500, 600, 700, 800 |
| 200 | 1000, 1200 |
| 400 | 1600, 2000 |
| 500 | 2500 |
| 1000 | 3000, 4000, 5000, 6000 |

1 A, 3 A, 6 A, 10 A, and 601 A are additional standard ratings for fuses.

## COMMON ELECTRICAL INSULATIONS

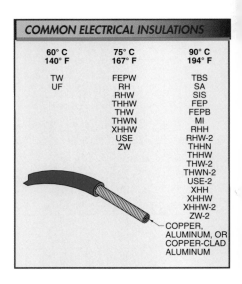

| 60° C 140° F | 75° C 167° F | 90° C 194° F |
|--------------|--------------|--------------|
| TW | FEPW | TBS |
| UF | RH | SA |
| | RHW | SIS |
| | THHW | FEP |
| | THW | FEPB |
| | THWN | MI |
| | XHHW | RHH |
| | USE | RHW-2 |
| | ZW | THHN |
| | | THHW |
| | | THW-2 |
| | | THWN-2 |
| | | USE-2 |
| | | XHH |
| | | XHHW |
| | | XHHW-2 |
| | | ZW-2 |

COPPER, ALUMINUM, OR COPPER-CLAD ALUMINUM

## NEMA Enclosure Selection

| Type | Use | Service Conditions | Tests | Comments | Type |
|------|-----|--------------------|-------|----------|------|
| 1 | Indoor | No unusual | Rod entry, rust resistance | | |
| 3 | Outdoor | Windblown dust, rain, sleet, and ice on enclosure | Rain, external icing, dust, and rust resistance | Do not provide protection against internal condensation or internal icing | 1 |
| 3R | Outdoor | Falling rain and ice on enclosure | Rod entry, rain, external icing, and rust resistance | Do not provide protection against dust, internal condensation, or internal icing | |
| 4 | Indoor/outdoor | Windblown dust and rain, splashing water, hose-directed water, and ice on enclosure | Hosedown, external icing, and rust resistance | Do not provide protection against internal condensation or internal icing | 4 |
| 4X | Indoor/outdoor | Corrosion, windblown dust and rain, splashing water, hose-directed water, and ice on enclosure | Hosedown, external icing, and corrosion resistance | Do not provide protection against internal condensation or internal icing | 4X |
| 6 | Indoor/outdoor | Occasional temporary submersion at a limited depth | | | |
| 6P | Indoor/outdoor | Prolonged submersion at a limited depth | | | |
| 7 | Indoor location classified as Class I, Groups A, B, C, or D, as defined in the NEC® | Withstand and contain an internal explosion of specified gases, sufficiently contain an explosion so an explosive gas-air mixture in the atmosphere is not ignited | Explosion, hydrostatic, and temperature | Enclosed heat-generating devices shall not cause external surfaces to reach temperatures capable of igniting explosive gas-air mixtures in the atmosphere | 7 |
| 9 | Indoor location classified as Class II, Groups E or G, as defined in the NEC® | Dust | Dust penetration, temperature, and gasket aging | Enclosed heat-generating devices shall not cause external surfaces to reach temperatures capable of igniting explosive gas-air mixtures in the atmosphere | 9 |
| 12 | Indoor | Dust, falling dirt, and dripping noncorrosive liquids | Drip, dust, and rust resistance | Do not provide protection against internal condensation | 12 |
| 13 | Indoor | Dust, spraying water, oil, and noncorrosive coolant | Oil explosion and rust resistance | Do not provide protection against internal condensation | |

## E

**equipment grounding conductor (EGC):** An electrical conductor that provides a low-impedance grounding path between electrical equipment and enclosures within a distribution system.

## F

**fault current:** Any current that travels a path other than the normal operating path for which a system was designed.

## G

**ghost voltage reading:** A reading on a voltmeter that is not connected to an energized circuit.

**grounded conductor:** A conductor that has been intentionally grounded.

**grounding:** A low-resistance conducting connection between electrical circuits, equipment, and the earth.

**grounding electrode conductor (GEC):** A conductor that connects grounded parts of a power distribution system (equipment grounding conductors, grounded conductors, and all metal parts) to an approved grounding system.

**grounding electrode system:** The connection of an electrical system to earth ground by using grounding electrodes, such as the metal frame of a building, concrete-encased electrodes, a ground ring, or other approved grounding method.

**ground loop:** An electrical circuit that has more than one grounding point connected to earth ground, with a voltage potential difference between the grounding points high enough to produce a circulating current in the grounding system.

## L

**leakage current:** Current that is not functional, including current in earth conductors and enclosures.

**low-impedance ground:** A grounding path that contains very little resistance to the flow of fault current to ground.

## M

**main bonding jumper (MBJ):** A connection in a service panel that connects the equipment grounding conductor (EGC), the grounding electrode conductor (GEC), and the grounded conductor (neutral conductor).

## O

**Ohm's law:** A mathematical formula stating that the current in an electrical circuit is directly proportional to the voltage and inversely proportional to the resistance.

## S

**separately derived system (SDS):**
An electrical system that supplies electrical power derived or taken from transformers, storage batteries, photovoltaic systems, wind turbines, or generators.

## T

**thermal imager:** A device that detects heat patterns in the infrared-wavelength spectrum without making direct contact with equipment.

**troubleshooting:** The systematic diagnosis of a system to locate any fault or problem.

*Page numbers in italic refer to figures.*

## C

circuit ground-fault current, 42
circulating current, 41
clamp-on ammeters, 49, 51, *52*
clamp-on ground testers, 49, 51, *52*
connections, 45–50, *47*, *50*
current circulation, 41

## D

digital multimeters (DMMs), 46, 49

## E

earth (ground) testers, 28
EGCs (equipment grounding conductors), 39
electrical test instruments, 51, *52*
electrodes, 27, 27–34
electronic equipment, 3–6
equipment grounding conductors (EGCs), 39

## F

fault current, 3, 5–6
four-pole testing, 15, *21*, 20–22
four-terminal ground resistivity measurements, *13*

## G

GECs (grounding electrode conductors), 7, *8*
ghost voltage readings, 47
ground clamp meters, *2*
ground current, 51, 52, 53

grounded conductors, 7
grounding, 1–2, 7–8
grounding categories, 5
grounding electrode conductors (GECs), 7, *8*
grounding electrode installation, 36–38, *37*, *38*
grounding electrode systems, 6
grounding systems, 3–8, *4*
ground-loop problems, 39–41, *41*
ground loops, *41*, 41
ground resistance values, 6–7
ground rod electrodes, 27, 32
ground system current, 53, *54*

## H

high ground current, 51–53

## I

improper ground connections, 46–47, *47*
insulation breakdown, 51–53
insulation testers, 51, *52*

## L

leakage current, 42, *44*
loose connections, 45–46, *47*
low-impedance grounding, 6

## M

main bonding jumpers (MBJs), 39, *40*
megohmmeters, 51
methods of grounding, 7–8
multiple grounding electrodes, 37–38